UNIFIED BEHAVIOR FRAMEWORK

FOR

DISCRETE EVENT SIMULATION SYSTEMS

I. Introduction

Modeling and simulation provides the capability to study, experiment, and research lifelike scenarios and systems without having to create and test those scenarios or systems in real life. In the case of scenarios that contain human actors, intelligent agents often represent the lifelike decision making of humans. Representing a lifelike scenario in a simulation system requires ever increasing scenario complexity, and as scenarios increase in complexity so too do the agents involved. With ever increasing complex tasks and situations, developers must devote more time to the agent's construction and should not be hindered by the constraints of the architecture being used for agent design. Different approaches to agent design offer developers different advantages to help combat these effects.

Cognitive design approaches focus around building a symbolic world model and reasoning over it to develop an action plan, which is computationally expensive and results in poor performance in dynamic environments [9]. In contrast to this are reactive control architectures which use a collection of behaviors that respond directly to stimulus from the surrounding environment. This in combination with some form of control-flow mechanism for action selection leads to robust performance in a changing environment. However reactive control architectures are still limited by their lack of planning for completion of high level goals.

1

Tiered and hybrid architectures solve this problem by taking advantage of the strengths of both approaches by merging them, using behaviors at a low-level for control of the agent, and deliberation and planning at a high-level for goal completion [9]. A disadvantage with behaviors that remains with a tiered approach is that behaviors at the low-level are typically tailored for a specific environment and situation, resulting in little code re-use for a different project or experiment. Furthermore in the case of a simulation framework, it does not allow for the flexibility of implementing different intelligent agent models.

1.1 Research Goal

Through its software engineering approach to intelligent agent design, the UBF has been identified as an effective means to address these concerns of flexibility and code re-use in intelligent agent design, however the UBF has yet to be used in a simulation environment. The goal of this research is to adapt the UBF to a discrete event simulation framework, specifically the AFSIM environment, which currently only uses Behavior Trees (BTs) for agent design. The additional objectives that come with this goal are to improve the overall agent design capabilities of AFSIM in relation to software engineering principles. Specifically this includes: 1) providing increased agent design flexibility to the developer, allowing them more freedom in choosing an agent architecture, 2) improving AFSIM's scalability through reduced code complexity, 3) simplifying development and testing through abstraction, and 4) promoting code reuse through a modular design.

1.2 Sponsor

This research is sponsored by the Aerospace Systems Directorate of the Air Force Research Laboratories (AFRL/RQQD) at Wright-Patterson Air Force Base. AFRL/RQQD requires a modeling and simulation framework for research of aerospace vehicles technology. Currently AFSIM is their current chosen modeling and simulation framework.

2

The work presented in this thesis provides necessary improvements to the intelligent agent design capabilities of AFSIM.

1.3 Assumptions

The terms and approaches used in this thesis are programming independent, however it is assumed that the reader has a basic knowledge of Object Oriented (OO) programming concepts and fundamental principles [10]. These concepts aid in understanding the motive for utilizing the UBF, as well as the structure of the UBF. Also while the general concepts are programming language independent, specific implementation of the non-real-time version of the UBF was performed in C++ as AFSIM is a discrete event simulation framework written in C++. Thus basic knowledge of C/C++ is assumed when discussing implementation. A basic understanding of the Unified Modeling Language (UML) is also assumed, specifically in regards to UML notation as discussion of the UBF includes its depiction through a UML diagram.

1.4 Thesis Structure

This thesis is structured as follows: this chapter introduces the problem and the goals of the research. Chapter II presents an overview of various intelligent agent architectures, including the advantages and disadvantages associated with each, as well as a comparison summary to the UBF. Chapter III illustrates the entirety of the research, condensed into a publishable paper. Finally Chapter IV provides additional details on AFSIM, the scenario used to test functionality of the UBF, and the implementation of the UBF into AFSIM.

II.   Intelligent Agent Architectures Background

This chapter presents an overview of current research into intelligent agent architectures. Intelligent agent architectures all strive to develop agents that accomplish tasks and high level goals in dynamic or unsafe environments. However each architecture is unique and has its own benefits, as well as its inherent disadvantages. To understand these benefits and select an architecture for implementation first requires understanding of these unique architectures.

The material in this chapter has been broken into four sections. First an early approach to robotics and Artificial Intelligence (AI) research is introduced, in which the paradigm was to sense and form a symbolic world model to reason and develop plans over. The next section is an examination of reactive control architectures, followed by discussion of tiered and hybrid architectures which attempted to merge the early approach with reactive controllers. The final section covers research into intelligent agents that reside specifically in software environments, which has a large focus on game AI.

2.1   Sense-Plan-Act Approach

In developing an autonomous robot, the Sense-Plan-Act (SPA) approach was the focus of AI research for 30+ years until the mid-1980's [9]. This approach first uses sensors to sense the world around the robot and then next using that information to form a world model. The robot would then use this model in planning algorithms to reason and form its next actions, then execute those actions and repeat these steps. A visual depiction of this approach is in Figure 2.1 and as can be seen it is a very linear cycle with serial execution.

The SPA approach to robotics proved problematic for even relatively simple tasks, as the world we live in is dynamic [1]. After only a few or even a single action, the robot's model of the world would no longer be valid and it would have to reform its next actions

4

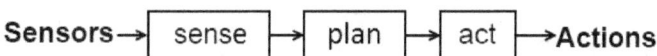

Figure 2.1: Robot control system using serial execution of functional modules [4].

to take. Forming the symbolic world model and reasoning over it is computationally very expensive, and thus having to perform these steps after every action, while simultaneously having no actions to execute, results in sub-par performance due to lag time between each action taken.

Using solely the symbolic world model for planning was an issue in itself as this meant the model must contain all information the robot needs [1]. Thus the robot can fail to perform in a dynamic environment where the robot would encounter foreign entities to its world model. The robot Shakey was an early example of the downfall to using this approach of forming a world model and planning around it [16].

2.2 Reactive Control Architectures

In response to the failures of using the SPA approach to developing intelligent agents, reactive control architectures (also referred to as behavior-based robotics) began to emerge [9]. These architectures focused on using the world as the model, as opposed to representing it symbolically, with the rational being that the world itself is its best representation. This results in a robot that responds to the world through behaviors as the robot senses the changing surrounding environment; visually this can be seen in Figure 2.2.

Of these emerging architectures, Braitenberg first presented a thought-experiment on developing intelligent agents that are composed of simple internal structures that react to the environment according to the structures' designs [3]. These simple structures can be thought of as behaviors as they define how to react to the environment. By slowly and

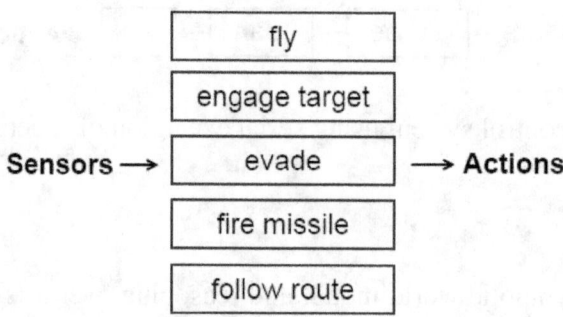

Figure 2.2: Robot control system using layered task achieving behaviors [4].

iteratively increasing the incorporation of these simple structures to develop a more and more complex intelligent agent eliciting ever more complex behavior, he demonstrated how an agent can be formed to solve any complex task using only simple behaviors.

Similarly Brooks argued that complex behavior of an agent is not the result of complex internal systems but instead the result of a complex environment, and thus the agent's internal systems should remain simple [4]. He also argued that traditional AI robots which model the world and then plan and act on that model must instead sense the world and act directly from that stimulus, in order to better achieve real-time interactions with the world. From this Brooks presented an architecture which linked the robots actions almost directly to its sensor stimulus.

2.2.1 Subsumption Architecture.

The Subsumption Architecture developed by Brooks incorporated parallel layers of competence, which can be thought of as task achieving behaviors, as opposed to the linear series of functional modules used in the SPA approach [4]. Each layer of competence is a sub-behavior of the complete behavior of the robot, and each higher level of competence includes within it the earlier levels of competence. There is no central control for the architecture, and thus each of these layers can be thought of as an individual agent operating asynchronously and independently in their own world, free to operate and output actions

6

to the robot as they find appropriate. Despite the lack of central control the layers work together, with the higher levels being able to subsume the lower levels, and the lower levels being able to inhibit the higher levels, resulting in viable actions being produced for the robot at all times. Furthermore with these layers all being independent of one another, the intelligence of the robot can be grown incrementally by simply adding on additional layers one layer at a time.

2.2.2 Motor Schema.

Another reactive control architecture that emerged around the same time was Arkin's Motor Schema, a schema-based reactive control [1]. A motor schema refers to a basic behavior such as "wander" or "avoid objects" which contains the knowledge required to generate a velocity vector command for that schema using potential fields. Thus the system uses sensors to channel environmental stimulus to the motor schema, and then each of the motor schema asynchronously generates these velocity vectors using solely that stimulus. The velocity vectors are then summed and normalized to be used for execution by the robot. In this manner, all behaviors are cooperating and contributing to the overall action produced by the robot. Because no persistent knowledge of the world is ever stored or used by the motor schema (unless the knowledge is known *a priori*), this architecture is truly reactive control and can encounter problems with local maxima/minima, as well as cyclic behavior.

2.2.3 Unified Behavior Framework.

Once an architecture is implemented onto a mobile robotic system for a specific application, it tends to bind that robot's capabilities to the architecture's strengths and weaknesses. Furthermore this leads to a platform specific model, as the behavior logic, the controller, and the underlying hardware all become interconnected. The UBF was developed as a means to combat this inflexibility associated with implementing behavior-based architectures, specifically at the controller level of the three-layer architecture [22, 23]. By separating the behaviors from the controller, different behavior-based systems

7

can be implemented into the behavior logic without having to change the controller. This also frees the behavior logic from being platform specific and encourages reuse.

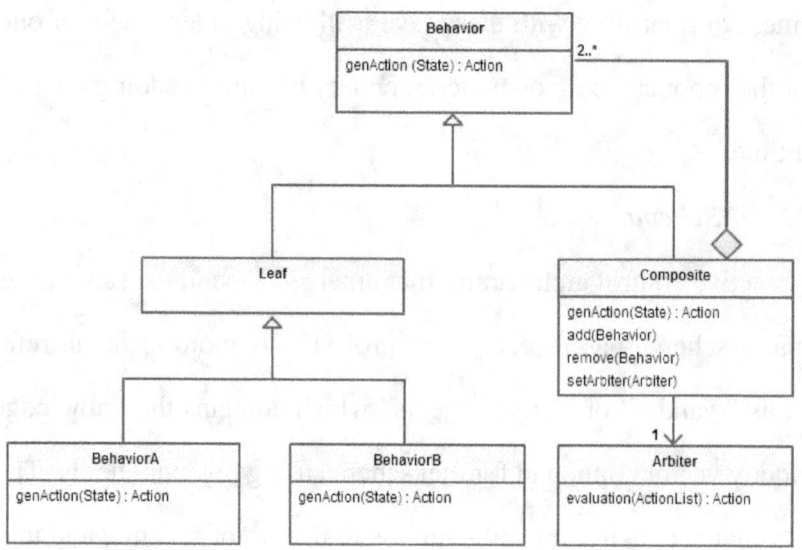

Figure 2.3: UML diagram for the UBF [22].

The UBF accomplishes this distinction through the software engineering principles of object oriented programming known as encapsulation, the strategy pattern, and the composite pattern [22], as shown with the UBF hierarchy in Figure 2.3. Behaviors are encapsulated as objects with an interface consisting of a generate action method (i.e., "*genAction()*") which returns an action object. Using this interface allows behaviors to be interchanged easily at runtime by the controller, following the strategy pattern. The *genAction()* method of the behavior uses a state object, which contains all perceived data from the agent's sensors, to produce the action object. The action object contains all necessary information (i.e. motor outputs) needed for the controller to execute the given action, as well as how strongly the behavior recommends the action to be taken (i.e. a vote amount set on an arbitrary scale). Thus the continuous cycle for the controller consists of first updating the state object with current sensor data, then next invoking the *genAction()*

method of the composite behavior so the behaviors generate a unified recommended action object, and then finally using this action object to execute commands to the agent.

Multiple behaviors can be encapsulated as a single composite behavior object, following the composite pattern. Composite behaviors can replace ordinary behaviors, allowing for limitless expansion of a UBF tree, and also encourages further modularity and reuse. Also contained within a composite behavior is a construct known as an arbiter. The arbiter is what unifies multiple action recommendations, given by the behaviors contained within the composite behavior, into a single action recommendation object. Arbiters can be customized to unify action recommendations however the developer sees fit, allowing for any behavior-based system implementation to be possible.

2.3 Tiered and Hybrid Architectures

Reactive control architectures were able to respond quickly in dynamic environments and as a result were viewed as a boon to robotic intelligence when compared to SPA-based robots [9]. However while these architectures had a quick response to their environment, they possessed limited planning or goal making capability. Tiered and hybrid architectures arose from the need to push through this capability ceiling, while still departing from the past SPA approach [2, 9]. Coincidentally multiple tiered architectures consisting of three-layers were inadvertently developed at the same time, each by different groups of researchers working relatively independent of one another. Despite being developed independently, these architectures all consist of three layers or components that are similar in their purpose and function.

2.3.1 Three-Layered Architectures.

Composition of a three-layer architecture is in general as follows [9]. At the low-level there is a component for reactive behaviors, at mid-level a sequencer that can swap out behaviors for the current situation at hand, and at the high-level a traditional deliberative planner. This can be seen visually in Figure 2.4. The low-level reactive component, called

9

the *skill layer* or *controller*, is a mechanism that implements the current skill-set or behavior scheme of the robot. These behavior schemes can represent tasks to be accomplished, and do so reactively by sensing the surrounding environment and reacting to it according to a transfer function, with no representation of the world needed (other than what the sensors provide) and should not rely on internal state. The *sequencing layer* or *sequencer* controls which behaviors should be currently executing, the purpose being that sequencing behaviors in the correct order allows for complex tasks to be accomplished. In order to know when to swap behaviors, the *sequencer* stores some amount of internal state to react to changes in the environment. Finally at the highest level is the *planning layer* or *deliberator*, which performs the typical planning aspect through polynomial-time and higher search algorithms.

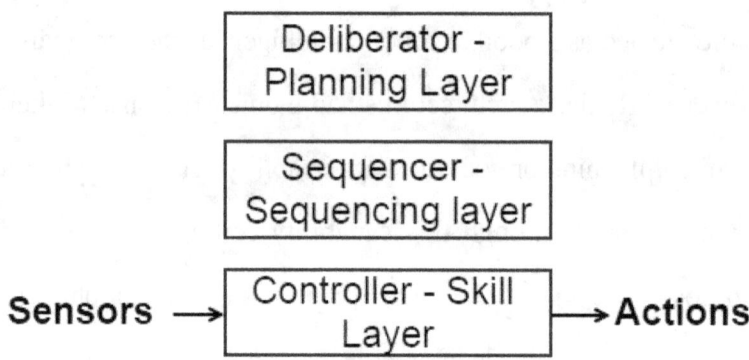

Figure 2.4: General composition for three-layered architecture.

Bonasso [2] and Gat [9] both developed similar three-layered architectures named 3T and ATLANTIS respectively which followed this general composition. A key difference between them was that ATLANTIS used more emphasis on the *sequencer* being in control, while 3T mostly used a *planner* in control approach. Worth noting is that 3T used the Reactive Action Packages (RAPs) [6] system developed by Firby for the sequencing capability of activating and deactivating various skillsets of the robot.

2.3.2 Saphira Architecture.

The Saphira architecture is an integrated sensing and control system for robotics applications [14]. It is similar to 3T and ATLANTIS in the sense of the general three-layer design, with reactive behaviors used at the low level, a planner available for query, and a sequencer in the middle to control behaviors, however there are some key differences. At the core is the Local Perceptual Space (LPS), which is a fusion of sensor and *a priori* information from independent perceptual routines which models the surrounding environment for the robot. The LPS is used by reactive behaviors to produce actions, and also by the Procedural Reasoning System (PRS) to sequence and monitor tasks for goal completion. Because the reactive behaviors and the PRS use the LPS in order to make decisions and produce actions, and because the LPS is a model of the world, then this model must be updated immediately and consistently so that the actions produced are valid even in a dynamic environment. Using Brooks' terms[4], this is accomplished with the overall organization of the architecture being partly horizontal and partly vertical, which can be seen visually in Figure 2.5. In this manner, higher behaviors and perceptual routines indicate a larger cognitive level of processing [14]. Thus there are perceptual routines at the low-level which quickly update the LPS using sensor information, and also perceptual routines at the high-level which require more processing time as they fuse sensor information with *a priori* information. Similarly higher level behaviors are more complex and are used to guide the lower level reactive behaviors.

2.3.3 POMDP-Based Layered Architecture.

Simmons' research in creating an autonomous mobile office delivery robot led to the development of the layered architecture, an architecture which is composed of five different layered architectures, and is based in using Partially Observable Markov Decision

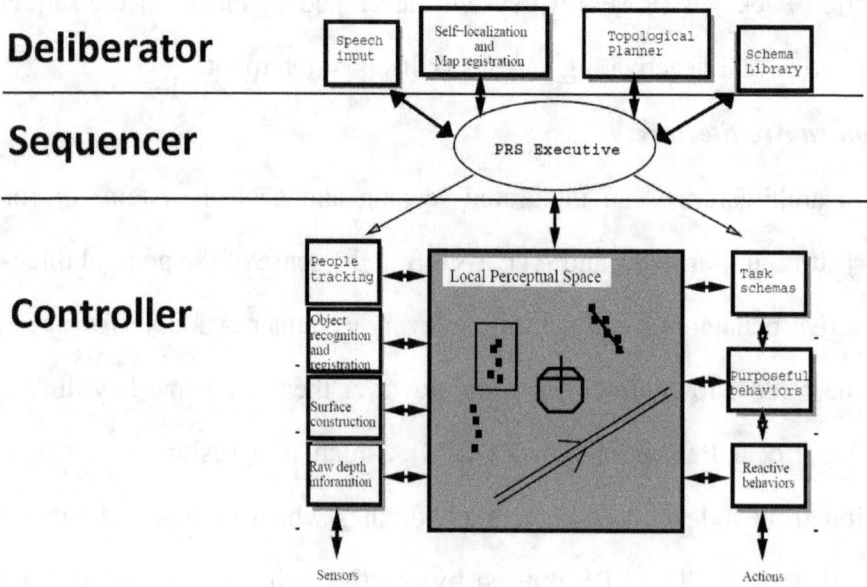

Figure 2.5: Visual depiction of the horizontal and vertical organization of the Saphira architecture. [14].

Process (POMDP) models for its navigation [13, 18]. At the bottom there is the servo-control layer which provides real-time motor control for the robot. Next up is the obstacle avoidance layer which uses an objective function (Simmons' Curvature-Velocity Method) that provides safety, speed, and progress along the desired heading while avoiding static and dynamic obstacles. From here is the POMDP-based navigation architecture which uses a POMDP model for a probability distribution on the robot's current location, which is then used by the robot for action decisions. After this is the path planning layer which decides how to travel from one location to another by choosing the path that is expected to have the highest utility from estimated travel time calculations. Finally is the multiple-task planning (task scheduling) layer, which consists of a symbolic non-linear planner (PRODIGY) to schedule delivery requests in the most efficient order in real-time as the requests arrive asynchronously.

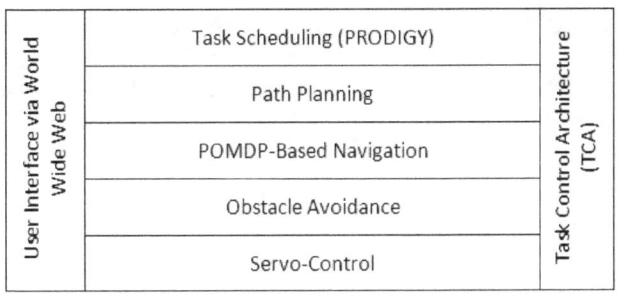

Figure 2.6: Layered Architecture [18].

As can be seen in Figure 2.6, these layers are all horizontal, distributed, and operating concurrently and thus must be integrated together [13, 18]. This is done using the Task Control Architecture (TCA) [19] which provides inter-process communication and synchronization, task decomposition and sequencing, execution monitoring, exception handling, and resource management. Finally to monitor progress and receive delivery requests from office workers, there is a user interface via the World Wide Web, providing live monitoring of task execution and command over the robot.

2.3.4 OpenR.

The Open Architecture for Robot Entertainment (OpenR) is an architecture that was developed particularly for autonomous robot systems in the entertainment field [8]. It was aimed at allowing robot developers the ability to build their own systems using Off-The-Shelf (OTS) components while meeting the specifications of OpenR. The specifications include a reference model consisting of a basic system, an extension system, and a development environment. The basic and extension systems revolve around using Configurable Physical Components (CPCs), which have common interfaces and Hot-Plug-and-Play capability to offer developers the flexibility to build their own robots by connecting these CPCs to the System Core. From here the model is also layered consisting of a Hardware Adaption Layer (HAL), a System Service Layer (SSL), and an Application Layer (APL), as can be seen in Figure 2.7. This layering allows developers to create

13

programs with user-friendly interfaces that are hardware independent. Further expanding upon the idea of common and well-defined interfaces, OpenR uses Apertos, an OO distributed real-time operating system which defines all components, both physical and software, as "objects" [24].

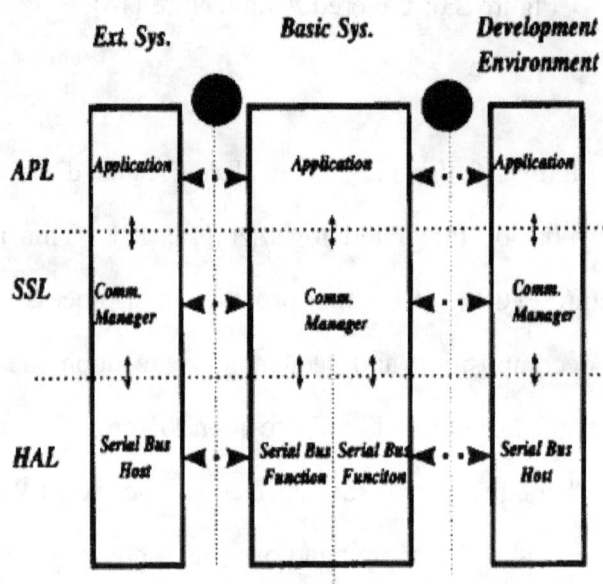

Figure 2.7: OpenR architecture's layered reference model [8].

2.4 Intelligent Agents in Software

In this section various architectures for developing intelligent agents in software are explored. Intelligent agents in software gave rise to new ways of developing autonomous agent architectures, largely in part due to the video game industry and its growing use of AI. The architectures discussed thus far are applicable to intelligent agents in all regards, however they were largely developed for use with robots and thus differ from these architectures and frameworks which focus solely on intelligent agents in software.

2.4.1 Behavior Trees.

A BT is a plan representation tool for action selection of an autonomous AI agent. BTs largely originated from AI development for Non-Player Characters (NPCs) for the video game Halo 2 [12], and as such BT frameworks are often game-oriented and are not generalized to other fields such as robotics [15]. However in the context of three-layered architectures, BTs reside in the middle or sequencing layer, where they provide a component for switching between the low-level reactive controllers.

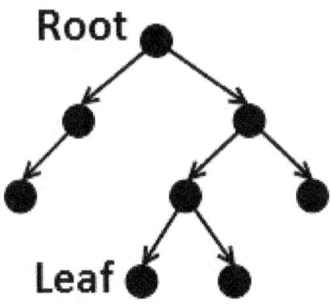

Figure 2.8: Graphical depiction of a rooted tree.

The structure of the BT is defined as a directed rooted tree, meaning it is a directed and connected acyclic graph with one node chosen as the *root*, as can be seen in Figure 2.8 [15]. A node incident to an outgoing edge is considered the *child* node of the connected pair of nodes, and a node with an edge leaving it is considered the *parent* node of the connected pair of nodes. Nodes which have a degree of only 1 are referred to as *leaf* nodes, with the exception of the unique *root* node. This collection of *leaf* nodes represents the behaviors of the agent which execute and produce an action. All other nodes of the tree are part of the mechanism for changing control-flow execution of these *leaf* nodes, and they consist of three different types, *selector*, *sequence*, and *parallel*.

Control-flow execution of the tree is as follows. The *root* node ticks an activation signal at a specified frequency rate f_{tick} which propagates down through the tree according

15

to the algorithms contained in the four different types listed above [15]. When this activation signal reaches a *leaf* node, the behavior contained in that node performs a check on whether it should execute or not. If it should execute, then it returns either *success* or *running*, however if the check fails, then it returns *failure*. This return value is then propagated back up the tree accordingly and control-flow execution of the tree continues until a return value reaches the *root* node, in which case the *root* waits for the appropriate time to tick another activation signal according to f_{tick}. An example of a fully constructed BT for a fighter jet agent participating in a sweep mission scenario is shown in Figure 2.9 [17].

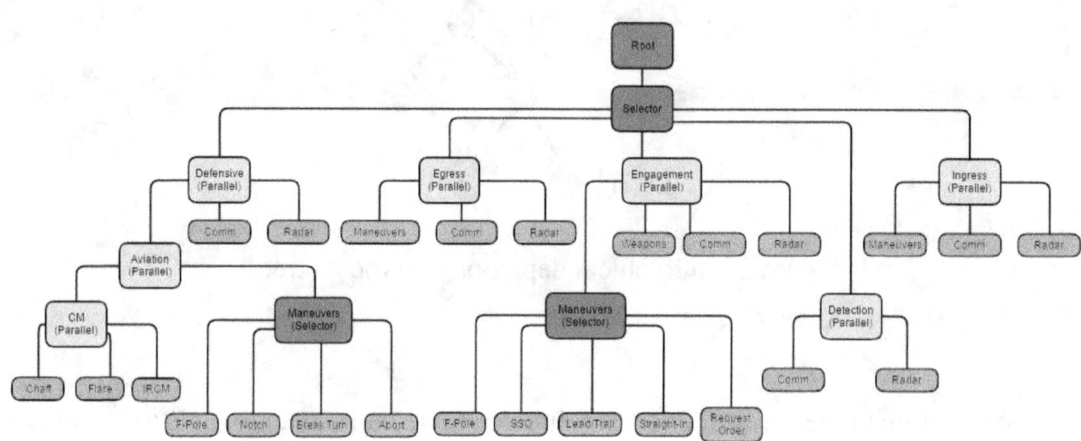

Figure 2.9: Example of a fighter jet agent's BT for a sweep mission scenario [17]. Note that *leaf* nodes are green, *parallel* nodes are yellow, *selector* nodes are blue, and the *root* node is red.

The algorithms for the three node types are as follows. A *selector* node ticks the activation signal sequentially across its children (from left to right if looking at a BT) until one of them returns either *success* or *running* [15]. If all children return *failure*, then the *selector* node returns *failure* as well, otherwise it returns either *success* or *running*, depending upon what the first non-failing child node returned.

16

A *sequence* node ticks the activation signal sequentially across its children (from left to right if looking at a BT) until one of them returns either *running* or *failure* [15]. If all children return *success*, then the *sequence* node returns *success* as well, otherwise it returns either *running* or *failure*, depending upon what the first non-success child node returned.

A *parallel* node ticks the activation signal sequentially across all of its children (from left to right if looking at a BT) [15]. If the number of children that return *success is* $\geq S$, S being a user-defined node parameter, then the *parallel* node returns *success*. If the number of children that return *failure* is $\geq F$, F being a user-defined node parameter, then the *parallel* node returns *failure*. If neither is true, then the *parallel* node returns *running*.

Algorithm 1: Selector [15]

1: **for** $i \leftarrow 1$ **to** N **do**
2:     *state* $\leftarrow Tick(child(i))$
3:     **if** *state = Running* **then**
4:         **return** *Running*
5:     **if** *state = Success* **then**
6:         **return** *Success*
7:     **end**
8: **end**
9: **return** *Failure*

Algorithm 2: Sequence [15]

1: **for** $i \leftarrow 1$ **to** N **do**
2:     *state* $\leftarrow Tick(child(i))$
3:     **if** *state = Running* **then**
4:         **return** *Running*
5:     **if** *state = Success* **then**
6:         **return** *Failure*
7:     **end**
8: **end**
9: **return** *Success*

Algorithm 3: Parallel [15]

1: **for** $i \leftarrow 1$ **to** N **do**
2:    $state_i \leftarrow Tick(child(i))$
3: **end**
4: **if** $\text{nSucc}(state) \geq S$ **then**
5:    **return** $Success$
6: **if** $\text{nFail}(state) \geq F$ **then**
7:    **return** $Failure$
8: **else**
9:    **return** $Running$
10: **end**

2.4.2 *Extended Teleo-Reactive Architecture: GRUE.*

Goal and Resource Using architecturE (GRUE) is an extended teleo-reactive architecture which uses resources to overcome inherent limitations of using a teleo-reactive architecture [11]. Teleo-Reactive Programs (TRPs) are composed of a list of rules, where each has a condition and an action. When the program is run the rules are evaluated and executed in series according to the first rule with a condition that evaluates to true. TRPs are each used to accomplish a single goal, and with multiple goals an arbitrator is used to determine with TRP should execute at each cycle. Goals are developed by the user, as well as the reward for accomplishing the goal, allowing for reinforcement learning. While TRPs have the advantage of handling a dynamic environment with ease due to arbitration, they also have limitations as all goals and their corresponding reward have to be set by the user, which is inappropriate for truly autonomous agents. GRUE solves these limitations by introducing the concept of a resource, which represents a condition that must be kept true throughout the execution of an action. By combining resources with priorities so that TRPs never use a resource required by a higher priority TRP, multiple goals and the appropriate reward become much easier to manage.

18

2.4.3 Dynamic Scripting.

Dynamic scripting is an online learning technique for game AI which builds upon scripting [20]. Scripting is a technique for implementing game AI which consists of lists of rules which are executed sequentially. While scripts are easy to understand and implement, they are generally long, complex, and static. This means that scripted AI are unable to adapt to unpredicted tactics from the opponent and that their weaknesses are easily exploitable. This results in a scripted AI whose tactics become completely predictable once the human understands his/her opponent. The only way for the developer to counter this problem when using scripted game AI is to extend the script and make it even larger and more complex, creating an endless challenge for the developer.

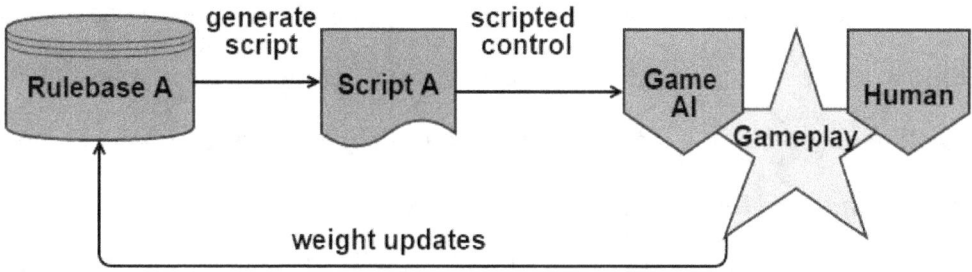

Figure 2.10: Dynamic scripting [20].

Dynamic scripting helps solve this problem by learning in real time how to win against the human player [20]. It does this by generating script from a weighted ruleset (with possible priorities). This script is then used to play against the human opponent and once an encounter has finished, the weights of the rules are adjusted according to how they contributed to the encounter so that the script generated for the next encounter is better prepared for the opponent depending on which tactics performed well. Figure 2.10 depicts this cycle visually. An important note with dynamic scripting is that the total weight is kept constant (ignoring remainder weight which is explained in the paper), and thus after

19

weight reward adjustments, all remaining weights must be adjusted to keep the total weight in check. Thus every rule weight is adjusted after every encounter, and then effectively all rules are always learning to some extent with every iteration.

2.4.4 SimBionic.

SimBionic is an AI modeling tool aimed at aiding behavior creation through an intuitive visual interface that is usable by both developers and non-programmers [7]. The tool uses Finite-State Machines (FSMs) as the underlying way to depict behaviors. Users define the actions and conditions of behaviors, i.e. the FSM's states and transitions, in the tool's graphical editor. Once these definitions have been made, they act as building blocks for behavior graphs to be constructed. The behavior graphs can then be indexed via a descriptor hierarchy to facilitate polymorphic behavior selection. There are also noteworthy extensions of SimBionic which augment its computational model: conditions are used to evaluate transitions (expression evaluation), a single state can refer to another FSM (stack-based execution), the capability to interrupt behaviors, and polymorphism.

2.4.5 Behavior Multi-Queues.

Behavior multi-queues [5] is an AI behavior architecture which uses behavior queues to fulfill five important traits that Cutumisu identified that a behavior architecture should display. The specified traits for behaviors include being responsive, interruptible, resumable, collaborative (joint behavior events), and generative (non-programmers are able to design and implement behaviors). The architecture encapsulates collections of behaviors as roles. Roles select behaviors which are made up of a sequence of tasks or other behaviors; using roles in this manner allows for the game AI to be responsive to a dynamic environment. When these behaviors are created, the sequence of tasks is placed in one of three queues depending on the behavior type. There is a queue for independent behaviors, a queue for collaborative behaviors, and a queue for latent behaviors (meaning they are event cued behaviors). Using roles consisting of behaviors for the game AI, in combination with multi-queues of this manner, allows for the architecture to fulfill all five traits. An example of the multi-queue architecture for a complex scenario can be seen in Figure 2.11.

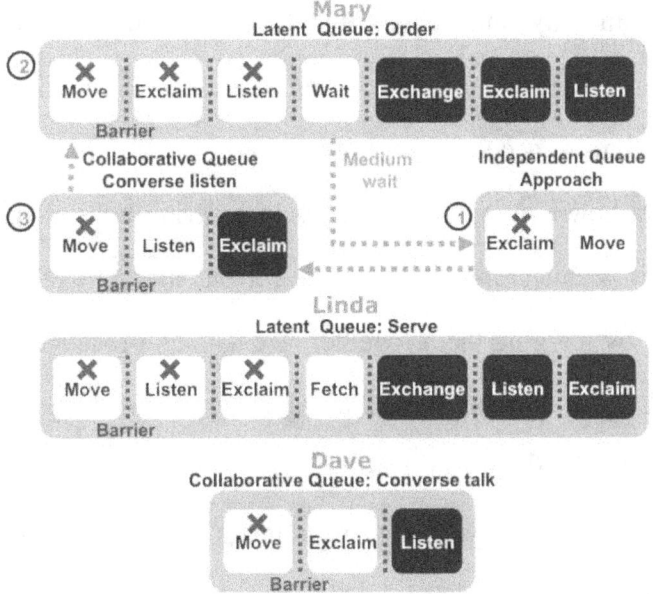

Figure 2.11: Complex scenario example depicted by Cutumisu using multi-queues architecture [5].

2.5   Choosing an Architecture - Summary

In summary, there are advantages and disadvantages to all of the software architectures presented. For example an architecture following the SPA approach can develop high level plans to accomplish goals, but is unresponsive while executing those plans and thus performance suffers in a dynamic environment. At the other end of the spectrum are reactive architectures that respond quickly even in a changing environment, but lack any type of forming of plans to accomplish high level goals, other than that which is inherent in the behaviors and how they are structured. Considering both of these approaches are the tiered and hybrid architectures which strive for the strengths of both of the SPA and reactive approaches by merging them together. By having a reactive/behavioral approach at the low level to control the motors, and a sequencer to utilize those behaviors in order to carry out plans, the deliberator can then form plans while still responding quickly to the world changing around it. However a disadvantage is that combining these layers for the aforementioned capabilities adds code base complexity and introduces potential interaction issues.

There are similar tradeoffs for architectures used to develop intelligent agents in software. Dynamic scripting builds upon scripting by correcting the challenge in complexity needed in order to make a script which adapts to the tactics of another opponents by changing its own tactics to become more effective. While it solves this issue, the drawbacks that are innate to scripting still exist. Dynamic scripting can be only be used when the AI agent can be scripted, which is not always possible. and there is also still a lack of complex collaboration and resumability between scripts. Dynamic scripting is also built towards improving its tactics, and thus lacks any inherent mechanism to improve its diversity, which can lead to static and predictable scripts [21].

SimBionic improved FSMs by adding in a more generative capability along with its four additional augmentations, however collaboration is still left to be desired. Similarly

BTs are much more intuitive to build and understand than their Hierarchical State Machine (HSM) and FSM predecessors, and also support reuse by making behaviors modular. However even with the recent additions of making BTs resumable and also cooperative[15], BTs still lack fusion between behaviors for more emergent behaviors of agents. Furthermore BTs can lead to agents which are slow to respond if behaviors early in the tree tick are computationally intensive. A natural evolution from BTs is behavior multi-queues, which allowed for all five of Cutumisu's desired features of a behavior to exist in a single architecture. However the lack of fusion between behaviors again has a lack of emergent behavior of agents as desired.

The UBF does not necessarily have any of these drawbacks, as it is a framework that offers the flexibility to implement or even merge any of the behavior-based systems mentioned. For example a UBF structure with the properly implemented arbiter can resemble and offer the same functionality as a BT, while also offering more capabilities such as fusion, which as an arbiter can be customized to emulate various algorithms [22]. Furthermore, as all of the behaviors in the UBF evaluate with each controller cycle, the downfall of slow response due to computationally intensive behaviors is mitigated. The UBF also promotes behavior reuse with its modular design and cross-platform code reuse. Performance degradation is a disadvantage that can result with the UBF if there are too many behaviors running concurrently on the agent and not enough computational resources to support the behaviors, as each behavior performs calculations with each controller cycle. Thus, as expected, as complexity of an agent grows, necessary computational resources grows as well.

III. Publishable Paper

This chapter presents the entirety of the research, illustrated as a publishable paper.

Unified Behavior Framework in Discrete Event Simulation Systems

Alex Kamrud*, Douglas Hodson†, Gilbert Peterson‡, and Brian Woolley§
Department of Electrical and Computer Engineering
Air Force Institute of Technology, Wright-Patterson AFB, OH 45433
Email: alexander.kamrud@afit.edu, douglas.hodson@afit.edu, gilbert.peterson@afit.edu, brian.woolley@ieee.org
Telephone: *651-485-8626, 937-255-3636 †x4719, ‡x4281, §x4618

Abstract—Intelligent agents provide simulations a means to add lifelike behavior in place of manned entities. Generally when developed, a single intelligent agent model is chosen, such as rule based, behavior trees, etc. This choice introduces restrictions into what behaviors agents can manifest, and can require significant testing in edge cases. This paper presents the use of the unified behavior framework (UBF) in the advanced framework for simulation, integration, and modeling (AFSIM) environment. The UBF provides the flexibility to implement any and all behavior-based systems, allowing the developer to choose the model he/she feels best fits the experiment at hand. Furthermore, the UBF demonstrates several key software engineering principles through its modular design, including scalability through reduced code complexity, simplified development and testing through abstraction, and the promotion of code reuse.

I. INTRODUCTION

The purpose of autonomous agents in simulation systems is to represent lifelike intelligence. In doing so, scenarios representative of real-life can be played out in a simulated environment for further study, experimentation, and research. To accomplish this, agents must be able to handle ever increasing complex tasks and situations, resulting in both their design and overall time in development growing as well. Different approaches to agent design offer developers different advantages to help combat these effects.

Cognitive design approaches focus around building a symbolic world model and reasoning over it to develop an action plan, which is computationally expensive and results in poor performance in dynamic environments. In contrast to this are reactive control architectures which use a collection of behaviors that respond directly to stimulus from the surrounding environment. This in combination with some form of control-flow mechanism for action selection leads to robust performance in a changing environment. However reactive control architectures are still limited by their lack of planning for completion of high level goals. Tiered and hybrid architectures solve this problem by taking advantage of the strengths of both approaches by merging them, using behaviors at a low-level for control of the agent, and deliberation and planning at a high-level for goal completion [1].

A disadvantage with behaviors that remains with a tiered approach is that behaviors at the low-level are typically tailored for a specific environment and situation, resulting in little code re-use for a different project or experiment. Furthermore

in the case of a simulation framework, it does not allow for the flexibility of implementing different intelligent agent models. To solve these issues, the unified behavior framework (UBF) has been identified as an effective means to developing cross-application behaviors which are re-usable due to modularity and drawing a delineation in design (i.e., a separation of concerns) between the behavior logic and the controller [2]. The UBF also provides flexibility to the user by allowing any agent model implementation to be possible. Finally the UBF also reduces code complexity and encourages experimentation through the composite pattern [3], allowing for new behaviors and control structures to be formed using any organization of existing behaviors and hierarchies.

This paper explores the use of the UBF to improve intelligent agent design in a discrete event simulation framework from a software engineering perspective. The remainder of this paper is structured as follows. In Section II background material related to approaches in developing intelligent agents are reviewed. Next a detailed overview of the UBF is given following the definition from [2], [4]. In Section IV results are presented for implementing the UBF into AFSIM using an attack on an integrated air defense system (IADS) scenario. Lastly conclusions and future work are presented.

II. BACKGROUND

A. Sense-Plan-Act Approach

In developing an intelligent agent, the sense-plan-act (SPA) approach was the focus of artificial intelligence (AI) research for 30+ years until the mid-1980's [1]. However the SPA approach to robotics proved problematic for even relatively simple tasks, as the world we live in is dynamic [5]. After only a few or even a single action, the robot's model of the world would no longer be valid and it would have to reform its next actions to execute. Forming the symbolic world model and reasoning over it to develop this action plan is computationally very expensive. Thus having to perform these steps after every action, while simultaneously having no actions to execute due to the dynamic environment invalidating the action plan, results in sub-par performance due to lag time between each action taken. The robot Shakey was an early example of the downfall to using this approach of forming a world model and planning around it [6]. Figure 1 depicts the linear SPA approach visually.

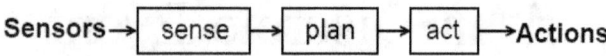

Fig. 1. Robot control system using serial execution of functional modules [7].

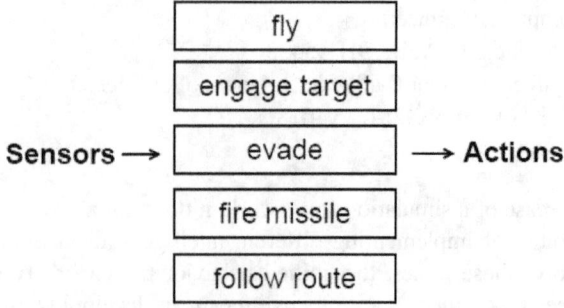

Fig. 2. Robot control system using layered task achieving behaviors [7].

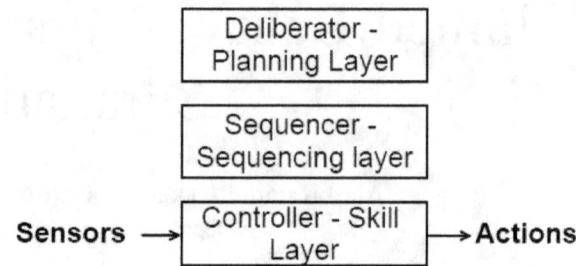

Fig. 3. General composition for three-layered architecture.

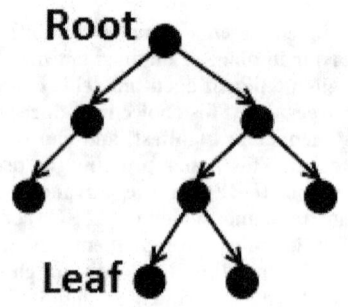

Fig. 4. Graphical depiction of a rooted tree.

B. Reactive Control Architectures

In response to the failures of using the sense-plan-act approach to developing intelligent agents, reactive control architectures (i.e. behavior-based robotics) began to emerge. These architectures focused on using low-level behaviors which responded directly to sensor stimulus [1]. Braitenberg first presented a thought-experiment on developing intelligent agents that are composed of simple internal structures that react to the environment according to the structures' designs [8]. By slowly and iteratively increasing the incorporation of these simple structures to develop a more and more complex intelligent agent eliciting ever more complex behavior, he demonstrated how an agent can be formed to solve any complex task using only simple behaviors. Similarly Brooks argued that complex behavior of an agent is not the result of complex internal systems but instead the result of a complex environment, and thus the agent's internal systems should remain simple [7]. From this, Brooks presented the Subsumption architecture, which linked the robot's actions almost directly to its sensor stimulus. The horizontal structure of this approach can be seen in Figure 2.

C. Tiered and Hybrid Architectures

Reactive control architectures were able to respond quickly in dynamic environments and as a result were viewed as a boon to robotic intelligence when compared to SPA-based robots [1]. However, while these architectures had a quick response to their environment, they lacked any type of planning or goal making, which represented a limitation in their capabilities. Three layer architectures and hybrid architectures arose from the need to push through this capability ceiling while still departing from the past SPA approach, and current research uses these architectures as the basis for their systems [1], [9]. In particular three layer architectures did this by using a reactive approach as a low-level control component and a

traditional deliberative planner at the high level, essentially merging the previous two approaches to building intelligent robots. This can be seen visually in Figure 3. The low-level reactive component, called the *skill layer* or *controller*, is a mechanism which implements the current reactive behavior scheme of the robot. In the middle is the *sequencing layer* or *sequencer* which controls the current active behavior scheme. Finally at the highest level is the *planning layer* or *deliberator*, which performs the typical planning aspect through polynomial-time and higher search algorithms.

D. Behavior Trees

Furthering research at the controller level is a construct known as a behavior tree (BT), and is what AFSIM currently uses to build intelligent agents. A BT is a plan representation tool for action selection of an autonomous agent. BTs largely originated from AI development for non-player characters (NPCs) for the video game Halo 2 [10], [11]. The structure of the BT is defined as a directed rooted tree, meaning it is a directed and connected acyclic graph with one node chosen as the *root*, as can be seen in Figure 4. A node incident to an outgoing edge is considered the *child* node of the connected pair of nodes, and a node with an edge leaving it is considered the *parent* node of the connected pair of nodes. Nodes which have a degree of only 1 are referred to as *leaf* nodes, with the exception of the unique *root* node. This collection of *leaf* nodes represents the different behaviors of the agent which produce and execute an action, meaning the behaviors perform the action execution of the agent. All other nodes of the tree are *branch* nodes and are part of the mechanism for

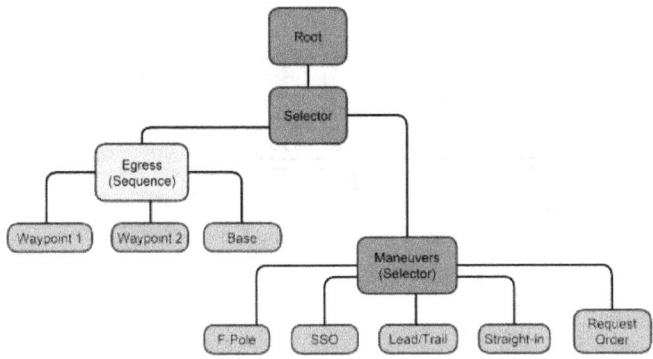

Fig. 5. Example of a fighter jet agent's BT for a sweep mission scenario [12]. Note that *leaf* nodes are green, *sequence* nodes are yellow, *selector* nodes are blue, and the *root* node is red.

changing control-flow execution of these *leaf* nodes. These *branch* nodes consist of various different types, however the *selector* and *sequence* nodes are two of the more common types that exist across all Behavior Tree implementations.

Control-flow execution of the tree is as follows. The *root* node ticks an activation signal at a specified frequency rate f_{tick} which propagates down through the tree according to the algorithms contained in the three different node types listed above [11]. When this activation signal reaches a *leaf* node, the behavior contained in that node performs a check on whether it should execute or not. If it should execute, then it does so for one cycle and returns either *success* or *running*. However if the check fails, then it returns *failure*. This return value is then propagated back up the tree accordingly and control-flow execution of the tree continues until a return value reaches the *root* node, in which case the *root* waits for the appropriate time to tick another activation signal according to f_{tick}. An example of a BT constructed for a fighter jet agent participating in a sweep mission scenario is shown in Figure 5 [12].

Algorithm 1: Selector [11]

1: **for** $i \leftarrow 1$ **to** N **do**
2:      $state \leftarrow Tick(child(i))$
3:      **if** $state = Running$ **then**
4:          **return** $Running$
5:      **if** $state = Success$ **then**
6:          **return** $Success$
7:      **end**
8: **end**
9: **return** $Failure$

To give examples of *branch* nodes, the algorithms for the *selector* and *sequence* node types are as follows.

1) Selector: A *selector* node ticks the activation signal sequentially across its children (from left to right if looking at a BT) until one of them returns either *success* or *running* [11]. If all children return *failure*, then the *selector* node returns *failure* as well, otherwise it returns either *success* or *running*, depending upon what the first non-failing child node returned.

Algorithm 2: Sequence [11]

1: **for** $i \leftarrow 1$ **to** N **do**
2:      $state \leftarrow Tick(child(i))$
3:      **if** $state = Running$ **then**
4:          **return** $Running$
5:      **if** $state = Success$ **then**
6:          **return** $Failure$
7:      **end**
8: **end**
9: **return** $Success$
---

2) Sequence: A *sequence* node ticks the activation signal sequentially across its children (from left to right if looking at a BT) until one of them returns either *running* or *failure* [11]. If all children return *success*, then the *sequence* node returns *success* as well, otherwise it returns either *running* or *failure*, depending upon what the first non-success child node returned.

BT's facilitate the construction and organization of a behavior scheme for an agent, and also give the user the advantage of re-use as each behavior is its own compact modular entity. However as BT's are implemented at the controller level of an agent and are thus interconnected in executing the agent's actions, the user is then unable to implement any other intelligent agent model and is instead tied to using solely BT's for agent representation. By incorporating the UBF, this paper further expands upon these benefits provided by BT's, while simultaneously allowing the user the flexibility to implement any intelligent agent model/architecture.

III. UNIFIED BEHAVIOR FRAMEWORK

Once an architecture is implemented onto a robotic system for a specific application, it tends to commit that robot's capabilities to the model's strengths and weaknesses. Furthermore this leads to an application specific model, as the behavior logic, the controller, and the underlying hardware all become interconnected. The UBF was developed as a means to address this inflexibility with implementing behavior-based architectures, specifically at the controller level of the three-layer architecture [2], [4]. By separating the behaviors from the controller, different architectures can be implemented into the behavior logic without having to change the controller. This also frees the behavior logic from being application specific and encourages reuse. The UBF accomplishes this distinction between behaviors and the controller through the software engineering principles of object-oriented programming known as encapsulation, the composite pattern, and the strategy pattern [2], as shown with the UBF hierarchy in Figure 6. Here in this section a description of the UBF is given following the definition from [2], [4].

A. Leaf Behaviors

As can be seen in Figure 6, there are two different concrete implementations of the abstract behavior interface (i.e. leaf and composite behaviors). Leaf behaviors contain a user-defined

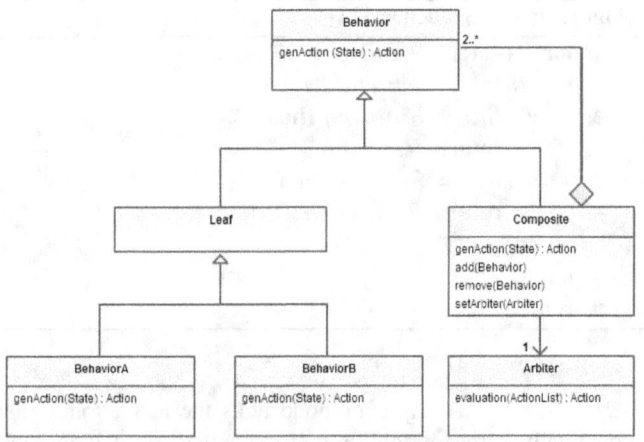

Fig. 6. UML diagram for the UBF[2].

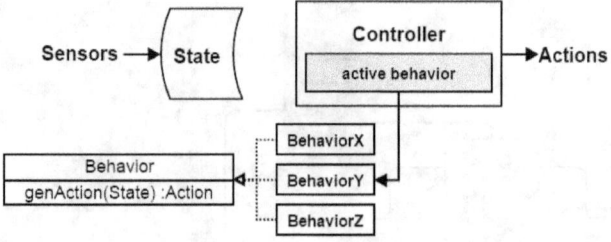

Fig. 7. Strategy pattern allowing for dynamic swapping of active behaviors [4].

"genAction" method necessary to generate a recommended action for the agent based on the current state perception of the agent. Encapsulated as an action object, this recommendation is composed of parameters necessary for the agent to execute the recommendation. There are also vote values for each parameter, and a separate vote value for the overall recommendation. These vote values indicate how strongly the behavior recommends the action or parameters be taken. By separating leaf and composite behaviors, the code complexity of action generation algorithms is confined to the leaf behaviors.

B. Composite Behaviors

Composite behaviors contain leaf or other composite behaviors as its children, and share the same interface as leaf behaviors, following the composite pattern [3]. By holding this set of sub-behaviors, $B = \{b_1, \dots, b_n\}$, the composite behavior generates a corresponding set of actions, $A = \{a_1, \dots, a_n\}$, by calling each child behavior's "genAction" method whenever the composite behavior's own "genAction" method is invoked. This allows the developer to treat both types of behaviors uniformly, and encourages reuse and experimentation as it provides the developer the ability to create new behavior structures using existing composite and/or leaf behaviors. Because composite behaviors extend the behavior interface in Figure 6, they must return a single action object as opposed to a set of actions. For this reason composite behaviors also contain a construct known as an arbiter which is used when the composite behavior is called to generate an action.

C. Arbiters

An arbiter uses a user-defined "evaluate" method to unify (i.e., select or fuse) all of the recommended actions in set A into a single action object. Arbiters can be customized to perform this action recommendation however the user sees fit, allowing for any architecture implementation to be possible. Composite behaviors can also use the "setArbiter" method to change its arbitration technique at any time, allowing for further modularity. Beyond an action's vote value and its associated parameters' vote values, an arbiter can also assign

weight values to each of the behaviors from a normalized set of weights $W = \{w_1, \dots, w_n\}$. By doing this, certain behaviors can be given priority in comparison to the other behaviors of the arbiter.

An example of an arbiter is an *overall vote value winner-takes-all (WTA)* arbiter. This arbiter operates by selecting the action with the highest overall vote value and returning it as the recommended action of the composite behavior. Weights for each individual behavior's vote value can also be used to tweak agent performance. The algorithm for this can be seen below in Algorithm 3. Take note that for this arbiter, if there are ties in the voting, then the sub-behaviors of the composite which are first called to generate actions will receive priority in selection.

Algorithm 3: WTA Overall Value - Evaluate ActionList

1: Action $winner = actionList[0]$
2: double $highestVote = winner.overallVote$
3: **for** all Action i in $actionList$
4:    **if** $i.overallVote > highestVote$
5:       $winner = i$
6:       $highestVote = i.overallVote$
7: **return** $winner$

D. Controller

The controller is responsible for querying its current active behavior for an action recommendation through the behavior's "genAction" method. Once it has the action object, the controller then executes this action, resulting in the agent performing actions in its environment [2]. As can be seen in Figure 7, the controller has the ability to hot-swap the current active behavior during execution due to the strategy pattern [3], as all behaviors share the same abstract behavior interface. Thus the controller is free to dynamically switch between any behavior-based architecture while the agent is running.

A typical sequence diagram for the controller is shown in Figure 8. Here the controller first updates the state object with the latest perception data from the agent's sensors. Then the controller sends this state object to the controller's current active behavior and invokes the behavior's "genAction" method. The behavior then uses the state information to produce an action recommendation object which is returned to the controller. Once the controller receives this action object,

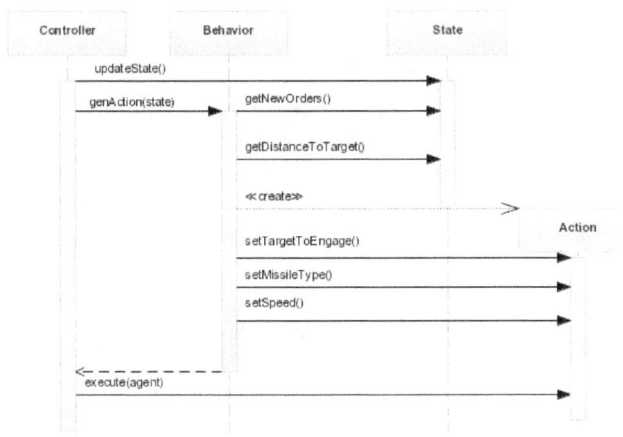

Fig. 8. Sequence diagram for the controller in the UBF [2].

the controller uses the action's "execute" method and uses the parameters within it to command the agent. Interesting to note that UBF behaviors are unable to act on the agent as BTs can. Rather, the agent explicitly applies the action recommended by the root behavior. While the agent is running, these steps repeat as a continuous cycle at a user-defined frequency rate.

IV. IMPLEMENTATION

In this section the UBF is implemented into a discrete event simulation framework to improve agent modeling capability in scenario simulations. Specifically this implementation occurs on the advanced framework for simulation, integration, and modeling (AFSIM).

A. AFSIM Background

AFSIM is a government-owned C++ simulation framework for use in constructing engagement and mission-level analytic simulations for the Operations Analysis community, as well as virtual experimentation. The primary goal of AFSIM applications is the assessment of new system concepts and designs with advanced capabilities not easily assessed with traditional engagement and mission level simulations. Development activities include modeling weapon kinematics, sensor systems, electronic warfare systems, communications networks, advanced tracking, correlation, and fusion algorithms, and automated tactics and battle management software. For external customization by the user (i.e. developing simulation scenarios), AFSIM provides a general-purpose scripting language which allows limited access to the framework using text input files (i.e. scripts). The scripting language is similar to Java/Visual Basic/C# and can be thought of as the "glue" which brings all components of the framework together for the user.

For design of autonomous agents, AFSIM uses a combination of BTs and the reactive integrated planning architecture (RIPR). RIPR is the framework included with AFSIM that enables behavior modeling for agents, and can be best thought of as a collection of utilities and algorithms that tie together nicely in the construction of an intelligent agent.

RIPR is mainly comprised of three components. There is the perception processor, the quantum tasker processor, and the actual behaviors of the agents. Behaviors for the agent are user-defined inside of script in AFSIM, and BTs used in conjunction with the user-defined behaviors allow for the modeling of the decision making of the agent.

The perception processor acts as the agent's cognitive model, and can be thought of as *state*. The agent senses the world by querying the platform and its subsystems for information. The agent builds knowledge internally, makes decisions, and then takes action by controlling its platform accordingly. Most platform queries and control actions take place inside of the AFSIM scripting language. As a RIPR agent maintains its own perception of threats, assets, and peers, this represents an agent's limited brain. To represent players of varying skill, each agent has its own tunable cognitive model. For example, an "expert" pilot agent can maintain knowledge of 16 threats that the agent updates (looks at radar) every 5 seconds.

The RIPR quantum tasker is used for commander subordinate interaction and task de-confliction. The quantum tasker comprises task generator(s), task-asset pair evaluator(s), an allocation algorithm, and various strategy settings (such as how to handle rejected task assignments). Each component (generator, evaluator, allocator) can be selected from predefined options, or custom created in script. It can send and/or receive tasks to/from other RIPR agents and other task manager platforms. The general cycle for the quantum tasker is for the generator to create tasks, the evaluator to consider task-asset pairings, and the allocator to then find the best asset for each task, which are then assigned accordingly.

Depending on the hardware/software used, some BT implementations do not allow for the simultaneous execution of different behaviors in the same agent. This limits the types of *branch* nodes that can be created, as they can then only tick their *child* behaviors one at a time. However, BTs inside of AFSIM do not suffer from this issue as AFSIM is a discrete event simulator, and thus has the infrastructure to allow for different pieces of script to execute and operate on an agent simultaneously. BTs that allow for this simultaneous execution of behaviors offer more variety in terms of *branch* nodes to the user, such as the *parallel* node.

A *parallel* node ticks the activation signal sequentially across all of its children (from left to right if looking at a BT) [11]. If the number of children that return *success* is $\geq S$, then the *parallel* node returns *success*. If the number of children that return *failure* is $\geq F$, then the *parallel* node returns *failure*. If neither is true, then the *parallel* node returns *running*. This algorithm can be seen below in Algorithm 4.

B. The UBF Implementation in AFSIM

Being that AFSIM is a simulation framework with behavior modeling tools already available for the user, some of the features of RIPR were used in conjunction with the implementation of the UBF. Specifically behaviors and their structured hierarchies are still written in AFSIM script by

Algorithm 4: Parallel [11]

```
 1: for i ← 1 to N do
 2:     state_i ← Tick(child(i))
 3: end
 4: if nSucc(state) ≥ S then
 5:     return Success
 6: if nFail(state) ≥ F then
 7:     return Failure
 8: else
 9:     return Running
10: end
```

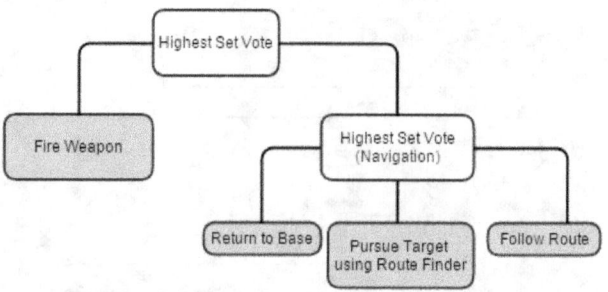

Fig. 9. Competitive approach using highest set vote arbiters.

users, the quantum tasker is still in use for commander-subordinate interaction, and the perception processor of RIPR is used as a suitable replacement to the state object of the UBF hierarchy. Arbiters are also still user-defined, however they are done so in C++ as user-defined methods of an arbiter object. Arbiter objects now also act as the composite behaviors, holding actions created by sub-behaviors, with a unified action being obtained by invoking one of the user-defined arbitration methods inside of the arbiter object. This still results in arbitration techniques being left to the design of the user, as well as retaining their modularity. The rest of the UBF was implemented as described in Section III.

For the purpose of this research AFSIM was used to simulate a Boeing-developed military combat scenario, specifically an attack on an IADS involving autonomous agents. In this scenario, four unmanned aerial vehicles (UAVs) with two standoff jammers (SOJs) carry out a deep-strike mission into the heart of an IADS with the objective of bombing six high-value targets. Inside the IADS the specific defenses for the targets include surface-to-air missile (SAM) sites, as well as four defensive fighter jets. Each fighter jet is an autonomous agent built with the UBF with the following behaviors/arbiters.

Behaviors:

Pursue Target using Route Finder - When at least one enemy target is available to the agent, this behavior activates and selects the highest priority target available to the agent and pursues that target utilizing AFSIM's route finder. The route finder allows an agent to path around static and/or dynamic obstacles using a depth-first-search algorithm to find the best route to the target while avoiding these obstacles. For this behavior it is used specifically to remain out of range of friendly SAM sites of the agent.

Return to Base - Calculates a direct route back to home base and flies that route at a pre-defined altitude and speed. This behavior activates when the agent's fuel amount dips below a certain threshold.

Fire Weapon - When an enemy target comes within firing

range of the agent's available weapons and the agent has a high enough track quality required for firing upon that threat, the agent selects the best weapon available and fires at the target.

Follow Route - This behavior is always active and causes the agent to fly on a pre-set patrol route with speed and altitude defined at discrete intervals along the route using waypoints.

Arbiters:

Highest Set Vote - This is a WTA arbiter that returns the action recommendation which has the highest overall vote value. This arbiter is used when the parameters of the action recommendations are to be used as a whole, as opposed to individually. Algorithm 3 contains the steps used for this arbiter and is considered a competitive approach, as behaviors compete against one another for selection of their respective action.

Activation Fusion - Action recommendations from all behaviors are fused together by selecting the individual parameters with the highest vote values and unifying those parameters into a single new action recommendation. Fusion arbiters such as this are best used when behaviors are designed to be cooperative and the behavior doesn't require use of all parameters to be successful. This algorithm is shown in Algorithm 5 and is a cooperative approach, as multiple behaviors can contribute to the final unified action.

Algorithm 5: Activation Fusion - Evaluate ActionList

```
1: Action unified = new Action
2: for all Action i in actionList
3:     if i.parameterXVote > unified.parameterXVote
4:         unified.parameterXVote = i.parameterXVote
5:         unified.parameterX = i.parameterX
6:     //***Repeat for all parameters***
7: return unified
```

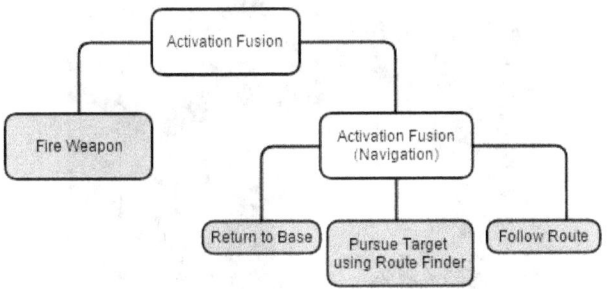

Fig. 10. Cooperative approach using activation fusion arbiters.

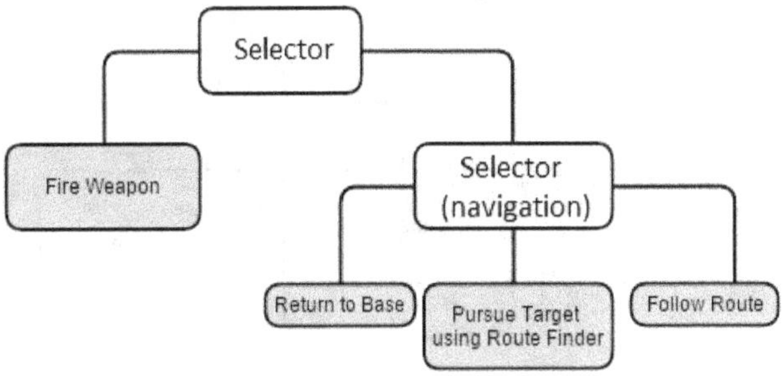

Fig. 11. Competitive approach agent built using BT.

C. Results

The simulation was executed using both a competitive approach as shown in Figure 9, and a cooperative approach as shown in Figure 10. Both simulations resulted in the SAM sites and autonomous agent defensive fighter jets successfully shooting down all four UAVs and the two SOJs. However there was a distinct difference between the two approaches in the speed to which the defenders reacted to the attack. The cooperative fusion approach reacted within 55.2 seconds to the attack, while the competitive approach reacted in 65.2 seconds, leading to a ten second difference between the approaches. This resulted in the cooperative approach shooting down the second UAV 13.9 seconds faster than the competitive approach. Ultimately this led to the fusion approach only losing three of the six high-value targets, while the competitive approach lost five of the six.

The performance gap is explained by the behaviors being constructed with a cooperative approach in mind. This means the behaviors were designed to achieve optimal performance by having their action recommendations unified together into a single action recommendation. Thus when used for a competitive approach, the behaviors suffer from disruption. Disruption occurs when the task the current active behavior is trying to accomplish is disrupted by a higher voting behavior, despite the higher voting behavior depending upon the lower voting behavior. An example of this is when the *Fire Weapon* behavior would interrupt the *Pursue Target* behavior to fire upon a target, despite the *Fire Weapon* behavior requiring an accurate track of the target which is provided by the *Pursue Target* behavior. The cooperative approach does not suffer from this as the action recommendations are unified together, allowing for the agent to fire upon the target while simultaneously pursuing it.

It is important to note that this does not imply one approach or architecture is superior to another. It only means that an intelligent agent model or architecture should be in mind from the beginning of behavior design. Had the behaviors been designed for a competitive approach, the outcomes could have certainly been reversed. The reason for implementing both approaches is to highlight the ability of the UBF to easily change an architecture or approach by leveraging the arbiters and modular hierarchy of the UBF. Switching between both approaches also requires minimal changes in code, demonstrat-

ing how the UBF reduces code complexity and constricts it to the leaf behaviors. Lastly implementing both approaches also exhibits code reuse as the same behaviors were used for both approaches, multiple experiments, and multiple agent platforms in simulation.

After the UBF was implemented inside the AFSIM framework, the majority of time spent in scenario development was in developing the individual leaf behaviors, as each leaf behavior is a unique piece of script with its own objectives. Once the leaf behaviors were developed, combining them to form different arbitration hierarchies was simple. Utilizing the arbiters as composite behaviors, the leaf behaviors could be conjoined into a single composite behavior with any arbitration technique selected. This technique could also be altered with ease as it only requires changing the arbitration method call to whichever was desired.

D. UBF versus BTs

Being that the UBF, BTs, and RIPR are all frameworks used to aid in the construction of an agent, the exact same agent can thus be constructed with any of these three frameworks, with no difference in performance between the agents. However where there is a difference is in ease of construction of the agent in terms of code complexity (lines of code). The competitive and cooperative agents constructed for the IADS scenario are used for these comparisons.

For the UBF versus BTs, in comparing the construction of an agent with verbatim decision-making to the competitive approach agent, the code complexity for both constructions is roughly equivalent. The reason for this can be seen in Figure 11, where the code for the leaf behaviors and behavior hierarchy for both the UBF and the BT is the same, with the only difference being a selector node substituted for a WTA arbiter and vice versa. However in comparing the construction of an agent with verbatim decision-making to the cooperative approach agent using the UBF versus BTs, the code complexity is greatly increased for the BT construction.

The reason for this is that without the simultaneous execution of leaf behaviors, actions which are always performed in conjunction with some other action must now be produced by every behavior, resulting in a significant increase in code complexity. Specifically in this example it means that the code for the *Fire Weapon* behavior must now be included with the

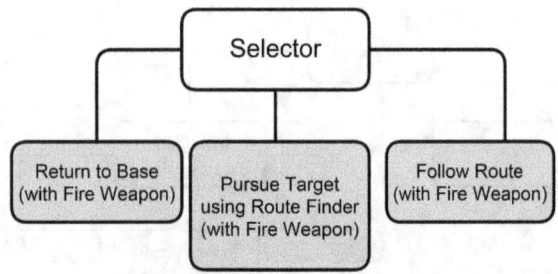

Fig. 12. Cooperative approach agent built using BT.

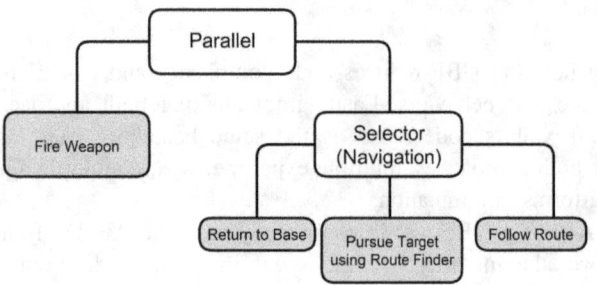

Fig. 13. Cooperative approach agent built using RIPR.

Fig. 14. Example scenario for illumination with evasion.

code of every other behavior, as *Fire Weapon* must be performed simultaneously while performing one of the navigation behaviors in order to verbatim replicate the decision making of the cooperative approach agent. Figure 12 contains the BT and illustrates this increase in code complexity. Thus the UBF has the advantage of reduced code complexity over BTs when constructing agents which require multiple behaviors executing simultaneously.

E. UBF versus RIPR

As RIPR uses BTs, the code complexity results of comparing the UBF versus RIPR in regards to the competitive approach agent are identical between the two, as the BT that RIPR uses for this agent is that which is shown in Figure 11. However for the cooperative approach agent, as RIPR has the ability to execute behaviors simultaneously, the BT for RIPR differs from that shown in Figure 12, and instead looks like the BT shown in Figure 13. As such, the code complexity of constructing the cooperative approach agent for the UBF versus RIPR does not differ, as their only difference in code is the arbiters/branch nodes making up the overall hierarchy. Thus there is no advantage between the UBF versus RIPR in regards to code complexity when constructing an agent in which both frameworks are capable of constructing.

However RIPR is unable to construct all agents that the UBF is capable of constructing. Specifically RIPR falls short of constructing any agent in which the agent requires the fusing of the action outputs of multiple behaviors. This is due to the fact that RIPR uses BTs for decision making, and with BTs the behavior logic is tightly coupled to the action execution of the agent, meaning the behaviors themselves perform the action execution of the agent. Thus there is no way to fuse

the action execution of different behaviors. As an example, if two different behaviors desired to navigate the agent in two different headings, there is no way for RIPR to fuse the headings and have the agent take a median heading between the two.

The UBF is capable of doing this as the behaviors in the UBF do not perform action execution for the agent and instead only produce an action object comprised of parameters, and these parameters are then able to be manipulated and fused together by the user-defined arbiters. Once these parameters are fused together as desired, the fused action object can then be executed, resulting in a blending of the action outputs of the behaviors. This added capability of agent construction in the UBF allows for greater experimentation and research, as this blending of action outputs can result in unforeseen emergent behaviors. This is further expanded upon in Section V. Additionally, while RIPR has the ability to construct an agent which replicates the same performance as a UBF agent utilizing the blending of action outputs, the resulting code complexity for the RIPR agent will be greater than that of the code complexity for the UBF agent.

This is shown with the following example scenario, where a blue fighter jet agent enters the state of having just launched a missile at a red target aircraft, and wants to keep the red target illuminated for additional missile guidance until detonation. However, the red target has also launched a missile at the blue fighter jet agent, so the blue agent wants to evade this oncoming missile, while also still simultaneously illuminating the red target. Figure 14 illustrates this scenario visually, and can be thought of as a "half-measure", as the agent isn't flying directly at the target for full illumination, but also not turning completely away from the target for full evasion. Furthermore, if the blue agent is not currently firing upon a target, the agent should perform full evasive maneuvers if an enemy launches a missile at the agent; and if the blue agent is not being targeted by an enemy and has a target to pursue or illuminate, then the blue agent should fully pursue that target.

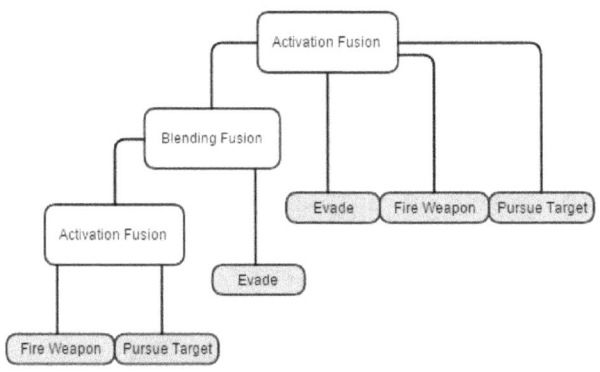

Fig. 15. UBF hierarchy for blue agent in example scenario.

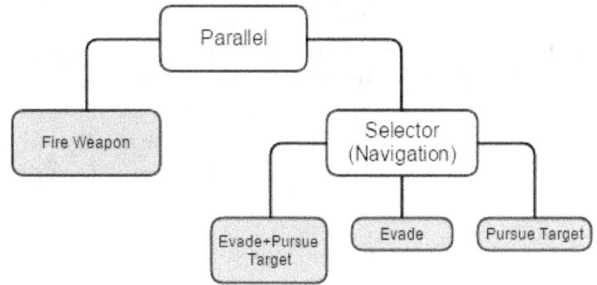

Fig. 16. BT for blue agent in example scenario.

To construct an agent in the UBF which replicates this performance, three behaviors are needed: an *Evade* behavior to evade an incoming missile, a *Fire Weapon* behavior to fire a missile, and a *Pursue Target* behavior to continue to pursue the target for beam illumination until missile detonation. The *Fire Weapon* and *Pursue Target* behaviors are combined using an Activation Fusion, allowing for the firing of a weapon on a target and continuous pursuit of the target until missile detonation. Then with the combined action from these two behaviors, and the action object of the *Evade* behavior, a Blending Fusion arbiter is used to blend the action outputs of these behaviors together, essentially averaging the parameters of the two action objects, creating a behavior hierarchy as shown in Figure 15. From this blending of the headings, a new emergent behavior is created, and the aircraft then flies as depicted in Figure 14. Note that the hierarchy accounts for evading enemy fire while not currently firing upon a target, by using an Activation Fusion arbiter in conjunction with the *Evade* behavior and the Blending Fusion arbiter, and likewise for pursuing a target while not being fired upon.

For RIPR to construct an agent which replicates this performance, a separate behavior strictly for simultaneous evasion and target illumination must be created, as RIPR is unable to blend the action outputs of the *Evade* and *Pursue Target* behaviors. Thus RIPR requires an additional *Evade+Pursue Target* behavior which performs the "half-measure" calculations; as well as a separate *Evade* behavior so that the agent can still perform full evasive maneuvers even while not firing upon a target, and a separate *Pursue Target* behavior so that the agent can still fully pursue a target while not being fired upon. Lastly the *Fire Weapon* behavior is needed for firing upon a target. This results in a BT as shown in Figure 16, and an increase in code complexity in comparison to the UBF, as an additional leaf behavior must be created in order to achieve the desired performance. If only three of these four behaviors are used to construct the RIPR agent, then agent performance will be lacking in terms of how the scenario describes the agent should perform. Thus, while RIPR can duplicate verbatim performance of a UBF agent which utilizes a Blending Fusion arbiter, the result is a RIPR agent with increased code complexity and subsequently more work for

the developer.

V. CONCLUSIONS

This paper demonstrates the benefits of using the UBF for constructing autonomous agents in a discrete event simulation system. Specifically the UBF reduces code complexity, simplifies development and testing, gives the flexibility to implement any agent model desired, and promotes code reuse through a modular design [2]. This was demonstrated by implementing a competitive approach with highest vote activation, and a cooperative fusion approach, where the actions of different behaviors were fused together. While behavior reuse and reduced code complexity are features that both the UBF and RIPR share, the additional advantage with the UBF is the flexibility to implement any behavior-based architecture into agent design. Without the UBF, AFSIM only offers RIPR as a means for control-flow execution of behaviors, leaving the user with only those node types mentioned in Section II & IV as mechanisms for designing control-flow of agent behaviors (Algorithms 1, 2, 4), and the user is forced to create behaviors which perform action execution of the agent. However with the UBF, agent construction has unlimited flexibility in behavior-based design as the behaviors solely output action objects as opposed to executing actions, and the arbiters that use these action objects are user-defined.

As mentioned in Section IV, this specifically gives the advantage of blending different behavior's action outputs, opening up more avenues for behavior experimentation of agents. Suggested future work would involve utilizing this blending of action outputs for research of optimal air-to-air tactics with future weapons systems. For example, optimal air-to-air tactics may be different for future UAV platforms, as UAVs do not have the constraint of the limitations of a human pilot. Thus the optimal action for a UAV in a certain situation may be different than current doctrine dictates. By defining a goal for the behaviors and letting them blend action outputs as needed, a better policy (tactics) for these future weapon systems could be found. This suggests that unforeseen emergent behaviors could aid in the discovery of unforeseen improvements in tactics.

ACKNOWLEDGMENT

The authors would like to acknowledge funding through the Aerospace Systems Directorate of the Air Force Research

Laboratories (AFRL/RQQD) at Wright-Patterson Air Force Base. The authors also gratefully acknowledge the support given by The Boeing Company, specifically developer Luke B. Miklos. The views expressed in this article are those of the author and do not reflect the official policy or position of the United States Air Force, Department of Defense, or the U.S. Government.

REFERENCES

[1] E. Gat, "Three-layer architectures," in *Artificial Intelligence and Mobile Robots*, D. Kertenkamp, R. Bonasso, and R. Murphy, Eds. AAAI Press, 1998, pp. 195–210.

[2] B. Woolley and G. Peterson, "Unified behavior framework for reactive robot control," *Journal of Intelligenct and Robotic Systems*, vol. 55, no. 2-3, pp. 155–176, July 2009.

[3] E. Freeman, E. Robson, B. Bates, and K. Sierra, *Head first design patterns*. " O'Reilly Media, Inc.", 2004.

[4] B. G. Woolley, G. L. Peterson, and J. T. Kresge, "Real-time behavior-based robot control," *Autonomous Robots*, vol. 30, no. 3, pp. 233–242, 2011.

[5] R. Arkin, "Survivable robotics systems: Reactive and homeostatic control," in *Robotics and Remote Systems for Hazardous Environments*, M. Jamishidi and P. Eicker, Eds. Prentice-Hall, 1993, pp. 135–154.

[6] N. J. Nilsson, "Shakey the robot," *SRI International*, 1984.

[7] R. Brooks, "A robust layered control system for a mobile robot," *IEEE Journal of Robotics and Automation*, vol. 2, no. 1, pp. 14–23, March 1986.

[8] V. Braitenberg, *Vehicles: Experiments in Synthetic Psychology*. Cambridge Massachusetts: MIT Press, 1984.

[9] R. Bonasso, D. Kortenkamp, D. Miller, and M. Slack, "Experiments with an architecture for intelligent reactive agents," in *Intelligent Agents II: Agent Theories, Architectures, and Languages*, M. Wooldridge, J. P. Mueller, and M. Tambe, Eds. Springer-Werlag, 1995.

[10] D. Isla, "Handling complexity in the halo 2 ai," in *Game Developers Conference*, vol. 12, 2005.

[11] A. Marzinotto, M. Colledanchise, C. Smith, and P. Ogren, "Towards a unified behavior trees framework for robot control," in *IEEE International Conference on Robots and Automation (ICRA 2014)*, 2014.

[12] S. L. G. G. L. Z. Sandeep S. Mulgund, Karen A. Harper, "Situation awareness for pilot-in-the-loop evaluation," Charles River Analytics, 725 Concord Ave. Cambridge, MA 02138, Tech. Rep. R96011, mar 1999.

IV. Additional Details

This chapter presents additional details on the research which were not covered in Chapter III. Section 4.1 presents some additional background on the AFSIM simulation environment. Section 4.2 discusses the scenario used for experimentation in greater detail. The final section presents how the UBF was adapted to the AFSIM environment in greater detail.

4.1   AFSIM Background

AFSIM is a government owned OO C++-based simulation environment that facilitates the rapid prototyping of customized engagement and mission-level warfare simulations. AFSIM includes a set of software libraries, shown as a functional architecture in Figure 4.1, containing routines commonly used to create analytic applications. The AFSIM infrastructure includes routines for the top-level control and management of the simulation; management of time and events within the simulation; management of terrain databases; general purpose math and coordinate transformation utilities; and support of standard simulation interfaces, such as those supporting the Distributed Interactive Simulation (DIS) protocol. The AFSIM component software routines support the definition of entities (platforms) to populate scenarios. These software routines contain models for a variety of user-defined movers, sensors, weapons, processors for defining system behavior and information flow, communications and track management. For external customization by the user (i.e. developing simulation scenarios), AFSIM provides a general-purpose scripting language which allows limited access to the framework using text input files (i.e. scripts). The scripting language is similar to Java/Visual Basic/C# and can be thought of as the "glue" which brings all components of the framework together for the user.

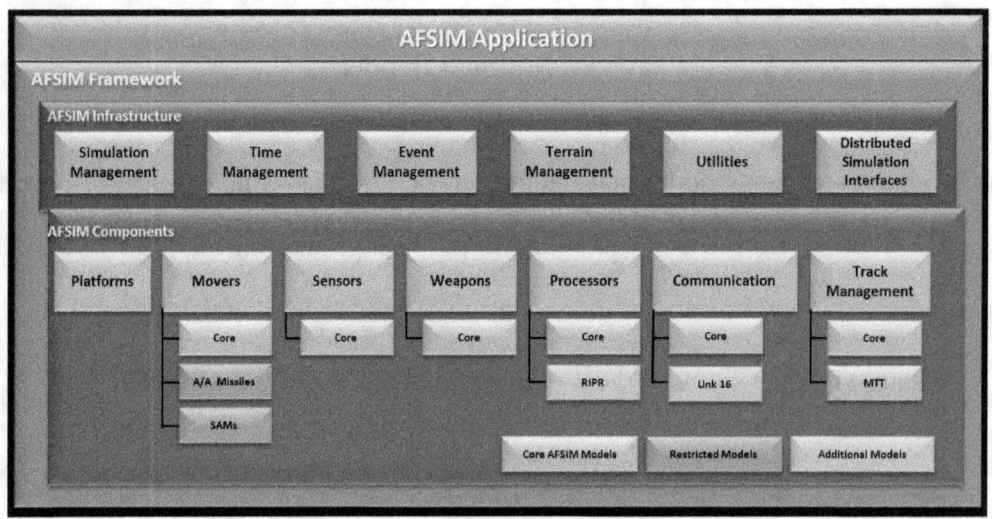

Figure 4.1: The AFSIM Functional Architecture.

The baseline AFSIM constructive application is called the Simulation of Autonomously Generated Entities (SAGE), and it is a simple application that reads in a user-defined input file of AFSIM script, executes the simulation, and outputs any user-defined data files. These output files are then used in the Visual Environment for Scenario Preparation and Analysis (VESPA) to visualize object positional time histories and other event information of the simulation. The Graphical User Interface (GUI) of VESPA can be seen in Figure 4.2. With SAGE the user can exercise all of the resident AFSIM capabilities, and the entire AFSIM IADS model is executed using the SAGE application. It is through recompiling this application that certain components of the UBF were incorporated into AFSIM.

4.2 IADS Scenario

As AFSIM was developed specifically to simulate threat Integrated Air Defense Systems (IADSs) to assess advanced air vehicle concepts performing Precision Engagement missions, an attack on an IADS scenario was deemed most appropriate for this research. In summary of the scenario, four Unmanned Aerial Vehicles (UAVs) with two Standoff

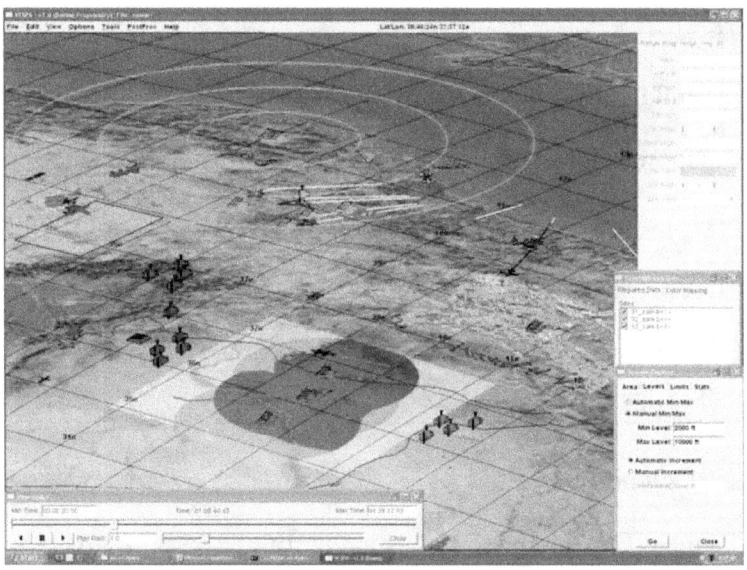

Figure 4.2: The VESPA GUI.

Jammers (SOJs) carry out a deep-strike mission into the heart of an IADS with the objective of bombing six high-value targets. Inside the IADS the specific defenses for the targets include a large radar company with Surface-to-air Missile (SAM) sites, and also four defensive fighter jets. Each fighter jet is an intelligent agent built with the UBF.

4.2.1 UAVs.

The UAVs are essentially dummy bombers with no AI involved. They each fly their own pre-determined ingress route over the targets, release ordnance once in relative range of the targets they are assigned to bomb, and then fly an egress route out of hostile territory and back to the ocean. They receive no orders and will only bomb the target they are assigned (UAV 1 assigned targets 1 and 2, UAV 2 assigned targets 1 and 3, UAV 3 assigned targets 4 and 5, and UAV 4 assigned target 6). Doing so allows for easier analysis of the results of the simulation which makes comparison of results between different intelligent approaches easier as well.

4.2.2 SOJs.

The SOJs original purpose was to support the UAVs as they ingress by flying a patrol route and suppressing enemy radar capabilities to limit detection of the UAVs. Unfortunately this capability was unable to be implemented as a portion of the Quantum Tasker required for this was missing in the source code released to AFIT. Instead of simply removing the SOJs, they were left in as they are still targets for the intelligent agents and thus serve a purpose in testing the agent's decision making.

4.2.3 Radar Company.

The radar company for the defenders is made up of a plethora of radars for detecting the enemy. There are nine Early-Warning Radar (EW) radars split into two separate squadrons, whose mission is detection of targets at long ranges and to then report it to the IADS commander. There is also two SAM battalions each consisting of an acquisition radar, a target-track radar, and five SAM launchers. These SAM battalions use the acquisition radar to scan their surrounding area and once a detection is made through the acquisition radar, the target track radar uses that information to obtain a more accurate lock on the target, which is then used to fire a SAM. As with the EW radars, this information is also passed up to the IADS commander. Through all of this the IADS commander receives these communications and uses it to assign tasks and tracks to its subordinates.

4.2.4 Fighter Jets - Intelligent Agents.

The four defensive fighter jets are intelligent agents designed using the UBF. At all times the fighters are using their behaviors to make intelligent decisions based upon their perceptions of the world, which includes target tracks, tasks given by commanders, and other sensor information. As the goal of these experiments is to confirm successful operation of the UBF, the perception processor is set at a "perfect pilot" setting, which removes the delay and info limit of the pilot's cognitive model. This was done as to ensure the perception processor is not interfering with differences in results between the different approaches in behavior design.

The fighters are split into groups of two, with each group reporting to a separate flight lead commander that is stationed on the ground. These flight leads communicate with the IADS commander in order to be alerted once targets are detected through the radar company. Once the flight leads are aware of these detections, the target tracks and tasks are passed down to the fighters. Once the fighters have this information they act upon it according to their behavioral design. The behaviors of these intelligent agents are listed below.

Behaviors:

Pursue Target using Route Finder - When at least one enemy target is available to the agent, this behavior activates and selects the highest priority target available to the agent and pursues that target utilizing AFSIM's route finder. The route finder allows an agent to path around static and/or dynamic obstacles using a depth-first-search algorithm to find the best route to the target while avoiding these obstacles. For this behavior it is used specifically to remain out of range of friendly SAM sites of the agent.

Return to Base - Calculates a direct route back to home base and flies that route at a predefined altitude and speed. This behavior activates when the agent's fuel amount dips below a certain threshold (1000 lbs of fuel, equaling approximately 142 seconds of flight time at a consumption rate of 7 lbs/second).

Fire Weapon - When an enemy target comes within firing range of the agent's available weapons and the agent has a high enough track quality required for firing upon that threat, the agent selects the best weapon available and fires at the target.

Follow Route - This behavior is always active (unless overridden by another navigation behaviors) and causes the agent to fly on a pre-set patrol route with speed and altitude defined at discrete intervals along the route using waypoints.

Arbiters:

Highest Set Vote - This is a WTA arbiter that returns the action recommendation which has the highest overall vote value. This arbiter is used when the parameters of the action recommendations are to be used as a whole, as opposed to individually. Algorithm 4 of Chapter III contains the steps used for this arbiter and is considered a competitive approach.

Activation Fusion - Action recommendations from all behaviors are fused together by selecting the individual parameters with the highest vote values and unifying those parameters into a single new action recommendation. Fusion arbiters such as this are best used when behaviors are designed to be cooperative and the behavior doesn't require use of all parameters to be successful. This algorithm is shown in Algorithm 5 of Chapter III and is a cooperative approach.

4.3 UBF Implementation in AFSIM

As described in Section III of Chapter III, the following components are necessary in order to successfully implement the UBF: 1) Action object, 2) State object, 3) The Controller, 4) Leaf Behaviors, 5) Arbiters, and 6) Composite Behaviors. The following subsections correspond to each of these components and discuss in further detail how each was implemented, and in some cases adapted, to the AFSIM environment, with reasons for these adaptations also being stated.

4.3.1 Action Object.

Action objects are implemented explicitly as described in Section III of Chapter III with no adaptation involved. Being directly built into the AFSIM framework, the user can create an action object inside of AFSIM script which will then hold various parameters required for action execution (e.g. altitude, speed, a route to fly, etc), and corresponding vote values required for arbitration. There is also an overall vote value that can be set by the user and is used to arbitrate between entire actions, and thus is used by WTA type arbiters. If this value is left unset by the user, then it defaults to the highest vote value associated with a parameter.

4.3.2 State Object.

The purpose of the state class is to store the agent's current perception of the world, usually updated through the agent's sensors. Then a state object can be sent to each behavior to be used for action generation. Behaviors can be still store their own behavior-specific state information if desired, however having a state class prevents each behavior from having to store all of its own state data, and thus avoids unnecessary duplication of data.

Currently AFSIM already accomplishes this through RIPR and its perception processor. The perception processor acts as the agent's cognitive model, using the sensors for that agent to update that agent's perception of the world. The behaviors of the agent can then access the agent's state in a manner similar to a blackboard system. Each agent has its own perception processor updating its own view of the world, and each agent's processor can be set to update at varying frequencies. As the perception processor is already implemented inside of AFSIM and can fulfill the role of the state class, it was adopted as substitution of this UBF component.

41

4.3.3   The Controller and Leaf Behaviors.

Currently in AFSIM behaviors are defined for use with BTs. As can be seen in Figure 4.3, behavior nodes are used to form a BT hierarchy. Each of these BT nodes corresponds to its own separate script file, which defines how the behavior operates. Each behavior will have a precondition portion of script which checks whether the behavior should execute or not (in accordance with how BTs operate as described in Section 2.4.1), and each will also have an execute portion of script which is what executes should the behavior pass the precondition. The only part of the behaviors which is relevant to the UBF is the execute portion, and this portion is defined solely in script. Thus leaf behaviors, which are responsible for action generation, can also be defined using AFSIM script.

```
processor  task_mgr   WSF_QUANTUM_TASKER_PROCESSOR
    update_interval 5 sec
    behavior tree
        selector
            behavior_node   evade
            behavior_node   go_home
            behavior_node   escort
            behavior_node   pincer_fsm
            behavior_node   pursue-target_route_finder
            behavior_node   planned_route
        end selector
        behavior_node   engage_weapon_task_target
    end behavior tree
```

Figure 4.3: Behavior Tree inside of an AFSIM script file.

To form a tree hierarchy for the UBF, AFSIM script can also be used. Each leaf behavior is defined in separate files of script, and each can be called upon similar to a method. Thus in combination with arbiters and some form of composite behaviors, a hierarchy can be built inside of script. This script file which contains the UBF hierarchy can also be responsible for executing the action generated by the tree, and thus be named the controller.

To set this up inside of AFSIM, a BT with a single behavior_node can be made and be aptly named "controller". This "behavior" is the only behavior of the BT, and always passes its precondition so that it always executes for the agent. By then incorporating the UBF hierarchy into the execute portion of this behavior through AFSIM script, the user is then implementing behavior logic into the agent in a manner in accordance with the UBF. The update portion of the controller for the UBF which controls the frequency at which the controller "ticks" can also be adjusted by changing the "update_interval" parameter of the BT, which can be seen at the top of Figure 4.3.

4.3.4   Arbiters and Composite Behaviors.

Arbiters are built directly into the AFSIM framework and thus an arbiter object can be created in script and used by the user. However as defining arbiters in script would prove difficult, arbiters differ from the traditional method described in Section 2.3.2 and instead were adapted to work with the AFSIM environment. Instead of having an arbiter interface which multiple different arbiters implement, the arbiter class built into the AFSIM framework instead defines all different types of arbiter techniques through user-defined C++ member functions. Thus once an arbiter object is created in script, any of these arbiter techniques can be called upon using an arbiter object and an action will be returned.

The arbiter techniques of these arbiter objects also differ from traditional implementation in that they no longer take in a list of actions, but instead the arbiter object itself holds the list of actions subsequently generated by child behaviors. In this manner the arbiter not only carries out arbitration, but also fulfills the role of composite behaviors. Leaf behaviors which generate an action object add it to their parent arbiter, and once the leaf/child behaviors of the arbiter have executed and given their action recommendations, the user can run any arbitration technique on the arbiter object and receive the unified recommended action for those child behaviors.